Assessment of Beddown Alternatives for the F-35

Executive Summary

Ronald G. McGarvey, James H. Bigelow, Gary James Briggs, Peter Buryk, Raymond E. Conley, John G. Drew, Perry Shameem Firoz, Julie Kim, Lance Menthe, S. Craig Moore, William W. Taylor, William A. Williams

RAND Project AIR FORCE

Prepared for the United States Air Force
Approved for public release; distribution unlimited

The research described in this report was sponsored by the United States Air Force under Contract FA7014-06-C-0001. Further information may be obtained from the Strategic Planning Division, Directorate of Plans, Hq USAF.

Library of Congress Cataloging-in-Publication Data

McGarvey, Ronald G.
Assessment of beddown alternatives for the F-35 : executive summary / Ronald G. McGarvey, James H. Bigelow, Gary James Briggs, Peter Buryk, Raymond E. Conley, John G. Drew, Perry Shameem Firoz, Julie Kim, Lance Menthe, S. Craig Moore, William W. Taylor, William A. Williams.
pages cm
Includes bibliographical references.
ISBN 978-0-8330-7808-7 (pbk. : alk. paper)
1. F-35 (Jet fighter plane)—Cost control. 2. United States. Air Force—Reorganization. 3. United States. Air Force—Appropriations and expenditures. I. Title.

UG1242.F5M397 2013
358.4'383—dc23

2013006083

Published 2013 by the RAND Corporation
1776 Main Street, P.O. Box 2138, Santa Monica, CA 90407-2138
1200 South Hayes Street, Arlington, VA 22202-5050
4570 Fifth Avenue, Suite 600, Pittsburgh, PA 15213-2665
RAND URL: http://www.rand.org/
To order RAND documents or to obtain additional information, contact
Distribution Services: Telephone: (310) 451-7002;
Fax: (310) 451-6915; Email: order@rand.org

Preface

As currently planned, the F-35 Joint Strike Fighter is the largest aircraft acquisition program in Department of Defense history. According to the December 2011 F-35 Selected Acquisition Report (DoD, 2011), the total acquisition cost to procure 2,457 F-35 aircraft across the United States Air Force (USAF), Navy, and Marine Corps is $331 billion, with total operating and support (O&S) costs of $617 billion to operate the aircraft through 2065 (both costs are computed using a base year of 2012). Moreover, the F-35 cost-per-flying-hour estimate has increased by more than 80 percent (in constant dollars) over the interval 2002 to 2010. To ensure that the affordability of the F-35 program is not threatened by continuing O&S cost growth, the USAF is examining alternative strategies to reduce those costs.

One approach to reducing O&S costs is to reduce the number of F-35 home-station operating locations, which is in turn related to the number of F-35 squadrons and the number of Primary Aerospace Vehicle Authorized (PAA) per squadron. In December 2011, RAND Project AIR FORCE (PAF) presented a preliminary analysis to the Director of Logistics, Office of the Deputy Chief of Staff for Logistics, Installations and Mission Support, Headquarters USAF (AF/A4L); this analysis found that increasing the combat-coded PAA per squadron while maintaining a constant total number of combat-coded PAA (thereby reducing the total number of squadrons) could significantly reduce maintenance manpower and support equipment costs.

Based upon these findings, the Vice Chief of Staff of the Air Force asked PAF to assess whether savings could be achieved by reconfiguring the USAF's 960 combat-coded F-35 PAA into larger squadrons (i.e., increasing PAA per squadron), by adjusting the PAA mix across the Active Component and Reserve Component, and by adjusting the percentage of Active Component PAA assigned to continental U.S. home-station locations. This report addresses how such changes would affect the USAF in the following ways:

- Ability to support both surge and steady-state contingency operations
- Ability to absorb the necessary number of F-35 pilots
- Requirements for maintenance manpower and support equipment
- Requirements for new infrastructure across the set of existing F-16 and A-10 bases
- Ability to develop future senior leaders out of the pool of fighter pilots.

This report provides an executive summary of our findings. It is a companion volume to the main report, which provides a more detailed presentation of our analytic methods:

> Ronald G. McGarvey et al., *Assessment of Beddown Alternatives for the F-35*, RR-124-AF, 2013.

This research was conducted within the Resource Management Program of RAND PAF for two fiscal year 2012 projects, "Reducing F-35 Operations and Sustainment Costs" and "Identifying Potential Efficiencies in the F-35 Basing Posture," sponsored at that time by, respectively, Major General Judith Fedder, AF/A4L, and General Philip Breedlove, the Vice Chief of Staff of the Air Force.

This report should be of interest to operations planners, logisticians, and manpower personnel throughout the USAF.

The views expressed are those of the authors, and do not necessarily reflect the official policy or position of the Department of the Air Force or the U.S. government.

RAND Project AIR FORCE

RAND Project AIR FORCE (PAF), a division of the RAND Corporation, is the U.S. Air Force's federally funded research and development center for studies and analyses. PAF provides the Air Force with independent analyses of policy alternatives affecting the development, employment, combat readiness, and support of current and future air, space, and cyber forces. Research is conducted in four programs: Force Modernization and Employment; Manpower, Personnel, and Training; Resource Management; and Strategy and Doctrine.

Additional information about PAF is available on our website:
http://www.rand.org/paf/

Acknowledgments

Many people inside and outside of the Air Force provided valuable assistance and support to our work.[1] We thank Major General Judith Fedder, AF/A4L, the original sponsor of our analysis of maintenance manpower and support equipment, along with Major General John Cooper, AF/A4L, our subsequent sponsor, for their support. We also thank General Philip Breedlove, the Vice Chief of Staff of the Air Force, for requesting that we expand these initial analyses of logistics resources to consider a broader set of USAF resources and requirements; and Major General Brett Williams, Director of Operations, Deputy Chief of Staff for Operations, Plans and Requirements, Headquarters U.S. Air Force (AF/A3O), for his assistance in defining the research directions and briefing our work to the F-35 Executive Review Board.

On their staffs, we extend a special thanks to Colonel Kyle Matyi, Lieutenant Colonel David Seitz, and Guy Fowl from AF/A4L; and to Colonel Jamie Crowhurst, Colonel James Jinnette, Colonel Steven Robinson, Lieutenant Colonel Brian Beales, Lieutenant Colonel Casey Tidgewell, and Major James Caplinger from AF/A3O for all of their assistance.

Staff from the Directorate of Studies and Analyses, Assessments and Lessons Learned, Headquarters U.S. Air Force (AF/A9) were extremely helpful in providing us with deployment requirements for the F-35; in particular we thank David Hickman, Maria Hayden and Michael Cusack for their assistance to our analysis.

This analysis brought together a Total Force team to ensure that the unique perspectives of the Active Component, Air Force Reserve Command (AFRC) and Air National Guard were represented. We received outstanding support from a very large number of participants at the Air Staff, Headquarters Air Combat Command, Headquarters AFRC, and the National Guard Bureau. We greatly appreciate the time and effort that all of the team members spent to provide us with the data—and the background understanding necessary to properly use those data—for our analysis.

At the Joint Program Office, we received assistance and data from Kimberly Fuller, Enass Saad-Pappas, and Sal Baglio. We thank Jennifer Bowles at the Air Force Cost Analysis Agency and Thomas Lies at the Directorate of Cost Analysis, Office of the Deputy Assistant Secretary of the Air Force for Cost and Economics (SAF/FMC), for providing us with cost data.

---

[1] All office symbols and military ranks are listed as of the time of this research.

At RAND, we thank our colleagues Laura Baldwin, Manuel Carrillo, Michael McGee, Patrick Mills, Carl Rhodes, and Al Robbert for sharing their insights and suggestions during the course of our analysis. We thank Robert Guffey for his assistance in developing our final project briefing, and are especially grateful to Kristin Leuschner, Megan McKeever, and Jane Siegel for their assistance in the preparation of this document. We thank our RAND colleagues John Ausink, Jeff Hagen and Steve Walters for their thorough reviews; their comments helped shape this report into its final, improved form.

That we received help and insights from those acknowledged above should not be taken to imply that they concur with the findings presented in this report. As always, the analysis and conclusions are solely the responsibility of the authors.

Assessment of Beddown Alternatives for the F-35: Executive Summary

As currently planned, the F-35 Joint Strike Fighter is the largest aircraft acquisition program in the Department of Defense's history. To ensure that the affordability of the F-35 program is not threatened by continuing operating and support (O&S) cost growth, the U.S. Air Force (USAF) is examining alternative strategies to reduce those costs. One approach to reducing O&S costs is to increase the number of Primary Aerospace Vehicles Authorized (PAA) per squadron, across a constant total number of USAF PAA, with a resulting reduction in the number of F-35 squadrons. In 2012, the commander of Air Combat Command (ACC/CC) approved a beddown plan to determine how to allocate the 960 combat-coded F-35 PAA across fighter squadrons and operating locations. The plan calls for 44 squadrons, with the aircraft allocated into squadrons of 24 PAA in the Active Component (AC) and Air Force Reserve Command (AFRC), and 18 PAA per squadron in the Air National Guard (ANG). These 44 total squadrons would be distributed among 31 operating locations.

At the request of the Vice Chief of Staff of the Air Force, RAND Project AIR FORCE (PAF) assessed whether O&S savings could be achieved by (1) reconfiguring the 960 combat-coded PAA into larger squadrons (i.e., increasing the PAA per squadron),[1] (2) adjusting the mix of PAA across the AC and Reserve Component (RC), and (3) adjusting the percentage of the AC PAA assigned to home-station locations in the continental United States (CONUS). This report addresses how such changes would affect the Air Force in the following ways:

- Ability to support both surge and steady-state contingency operations
- Ability to absorb the necessary number of F-35 pilots
- Requirements for maintenance manpower and support equipment (SE)
- Requirements for new infrastructure across the set of existing F-16 and A-10 bases
- Ability to develop future senior leaders out of the pool of fighter pilots.

[1] For USAF fighter aircraft, no current squadron has more than 24 PAA. However, fighter squadron sizes have varied over time based on the facilities and aircraft numbers available, and they tend to peak during wartime and decrease during postwar drawdown periods. The analysis presented in this report will examine the potential for squadron sizes larger than 24 PAA to generate increased cost-effectiveness.

Increasing Squadron Size Has the Potential for Cost Reductions

Our primary finding is that increasing the F-35 squadron size from the levels utilized in the ACC/CC-approved beddown (24 PAA per AC and AFRC squadron, 18 PAA per ANG squadron) can satisfy both expected surge and steady-steady deployment requirements and can generate significant savings in the following areas:

- Annual pilot absorption flying costs (more than $400 million)
- Annual maintenance manpower costs (more than $180 million)
- One-time support equipment requirements (more than $200 million)
- Annualized facilities costs (more than 10 percent).

The lower bounds on these estimates could be achieved, and all deployment requirements satisfied, if the USAF were to implement a posture that utilizes 30 PAA per AC and AFRC squadron and 24 PAA per ANG squadron (beddown alternatives 2G and 2H, in Table 1). The savings would increase if the USAF were to select a posture with 36 PAA in AC and AFRC squadrons and 24 PAA in ANG squadrons (Alternatives 2I and 2J), but this posture would assume increased risk; it has sufficient squadrons to satisfy surge wartime requirements, but it cannot satisfy steady-state requirements within the desired deploy-to-dwell ratios.

Further savings are possible in all categories except maintenance manpower, if the percentage of PAA in the AC were increased from the 60 percent level assumed in the ACC/CC-approved beddown. The percentage of AC PAA assigned to CONUS locations had little impact on these savings.

Set of Beddown Alternatives Considered

A set of 28 alternative beddowns was examined in this analysis. For the purposes of this report, beddown refers to the number and sizes of F-35 squadrons, and their distribution across AC CONUS, AC outside the continental United States (OCONUS), ANG and AFRC, without regard to the specific locations at which these squadrons are permanently based.[2] The full set of alternatives is presented in Table 1, which also introduces the naming convention that will be used throughout the remainder of this

---

[2] We recognize that this use of the term beddown is inconsistent with AFI 10-503, Strategic Basing, Washington, D.C., September 27, 2010, which states "Beddown is considered the execution of a basing action." We use the term beddown here in a different manner, and avoid the use of the term basing, to emphasize that this analysis is not focused on specific locations for permanently stationed F-35 units.

report.[3] These beddowns varied the squadron size between 24 and 36 PAA for AC and AFRC, and between 18 and 24 PAA for ANG.[4] These beddowns also varied the percentage of combat-coded PAA in the AC between 45 and 75 percent, and the percentage of total AC PAA in CONUS between 50 and 67 percent. The ACC/CC-approved beddown corresponds to beddown alternative 2A.

More detail regarding how this number of squadrons was arranged into multi-squadron wings for the AC is presented in Table 2. We will assume that each AFRC and ANG squadron is located at a unique base. We assume that each AC Wing, as presented in Table 2, corresponds to one base. For example, beddown 1A corresponds to 7 AFRC bases; 20 ANG bases; three AC CONUS bases, with three squadrons of 24 PAA at each base; and four AC OCONUS bases, one base with three squadrons of 24 PAA, and two squadrons of 24 PAA at each of the other three AC OCONUS bases.[5]

---

[3] Note that, under this structure, there are two beddown alternatives (3G and 3H) that are not included in this analysis. These are shaded in gray in Tables 1 and 2. We exclude them due to the difficulty of allocating 25 percent of the total combat-coded PAA to the RC with AFRC squadrons of 30 PAA and ANG squadrons of 24 PAA. The only option for which the arithmetic works has a total of 120 PAA in AFRC and 120 PAA in ANG, which is inconsistent with all other beddowns, for which ANG has many more PAA than does AFRC.

[4] The analyses presented in this report assume that each AFRC and ANG squadron is located at a single base. This assumption is consistent with the current beddown of combat-coded AFRC and ANG fighter/attack squadrons. It is possible that multiple RC squadrons could be assigned to a single wing at a single base, but this analysis did not consider such alternatives. This analysis does, however, examine the efficiencies associated with multi-squadron wings for AC squadrons.

[5] There are significant cost implications associated with changing the number of USAF bases at which fighter/attack aircraft are permanently stationed. Issues related to the closure or repurposing of existing USAF bases were beyond the scope of this analysis.

Table 1. F-35 Beddown Alternatives

Percent of Total PAA in AC	Squadron Size (PAA)	Percent of Total AC PAA in CONUS	Beddown Alternative	AC Squadrons		RC Squadrons		Total Squadr
				CONUS	OCONUS	AFRC	ANG	
45	18 (ANG), 24 (AC/AFRC)	50	1A	9	9	7	20	45
		67	1B	12	6	7	20	45
	24 (ANG), 24 (AC/AFRC)	50	1C	9	9	6	16	40
		67	1D	12	6	6	16	40
	18 (ANG), 30 (AC/AFRC)	50	1E	8	7	5	20	40
		67	1F	10	5	5	20	40
	24 (ANG), 30 (AC/AFRC)	50	1G	8	7	5	15	35
		67	1H	10	5	5	15	35
	24 (ANG), 36 (AC/AFRC)	50	1I	6	6	4	16	32
		67	1J	8	4	4	16	32
60	18 (ANG), 24 (AC/AFRC)	50	2A	12	12	4	16	44
		67	2B	16	8	4	16	44
	24 (ANG), 24 (AC/AFRC)	50	2C	12	12	4	12	40
		67	2D	16	8	4	12	40
	18 (ANG), 30 (AC/AFRC)	50	2E	10	9	4	15	38
		67	2F	13	6	4	15	38
	24 (ANG), 30 (AC/AFRC)	50	2G	10	9	5	10	34
		67	2H	13	6	5	10	34
	24 (ANG), 36 (AC/AFRC)	50	2I	8	8	4	10	30
		67	2J	11	5	4	10	30
75	18 (ANG), 24 (AC/AFRC)	50	3A	15	15	4	8	42
		67	3B	20	10	4	8	42
	24 (ANG), 24 (AC/AFRC)	50	3C	15	15	3	7	40
		67	3D	20	10	3	7	40
	18 (ANG), 30 (AC/AFRC)	50	3E	12	12	2	10	36
		67	3F	16	8	2	10	36
	24 (ANG), 30 (AC/AFRC)							
	24 (ANG), 36 (AC/AFRC)	50	3I	10	10	2	7	29
		67	3J	13	7	2	7	29

Table 2. Arrangement of AC Squadrons Into Multi-Squadron Wings, for Each Alternative Beddown

Beddown Alternative	AC Squadron Size (PAA)	Number of Wings with *N* Squadrons						
		AC CONUS				AC OCONUS		
		1	2	3	4	1	2	3
1A				3			3	1
1B	24			4		2	2	
1C				3			3	1
1D				4		2	2	
1E			1	2		1	3	
1F	30		2	2		3	1	
1G			1	2		1	3	
1H			2	2		3	1	
1I	36			2		2	2	
1J			1	2		4		
2A		1	1	3			6	
2B	24		2	4			4	
2C				4				4
2D			2	4			4	
2E			2	2			3	1
2F	30		2	3		2	2	
2G			2	2			3	1
2H			2	3		2	2	
2I	36		1	2			4	
2J			1	3		3	1	
3A				1	3			5
3B	24				5		2	2
3C				1	3			5
3D					5		2	2
3E	30			4				4
3F			2	4			4	
3G								
3H								
3I	36		2	2			2	2
3J			2	3		1	3	

An important aspect of this analysis is that it was directed to assume that all RC units and all AC units in CONUS would utilize associate unit arrangements. Thus, every beddown alternative includes both *Active Associate* units, in which an RC unit has principal responsibility for a weapon system and shares the equipment with an AC unit, and *Classic Associate* units, in which an AC unit retains principal responsibility for a weapon system and shares the equipment with an RC unit.

Findings

All 28 Beddown Alternatives Satisfy Surge Deployment Requirements

For surge, we assume that all combat-coded squadrons in both the AC and RC are available for tasking. A key assumption in this analysis was that each squadron contained one independent or "lead" Unit Type Code (UTC).[6] Thus, each squadron could deploy to and operate out of, at most, one location, regardless of squadron size, consistent with USAF policy for resourcing legacy fighter squadrons for deployment. Based on modified Integrated Security Construct (ISC) scenarios, provided by the Directorate of Studies & Analysis, Assessments and Lessons Learned, Headquarters U.S. Air Force (USAF/A9), we identified the number of squadrons that would need to deploy to satisfy surge requirements. These scenarios envision an employment construct in which the F-35 is deployed in a manner similar to the F-16. For campaign (i.e., warfight) activities, the squadron size has an impact on the number of squadrons that need to be deployed. This is because, in surge scenarios, it is not uncommon for a large number of aircraft to be deployed to one location. We found that all 28 beddown alternatives could satisfy surge requirements. Figure 1 demonstrates how the number of F-35 squadrons available in each of the 28 alternative beddowns compares to these requirements.

[6] A UTC is a unit of capability specified by the required manpower and equipment.

Figure 1. Ability of Alternative F-35 Beddowns to Satisfy Surge Deployment Requirements

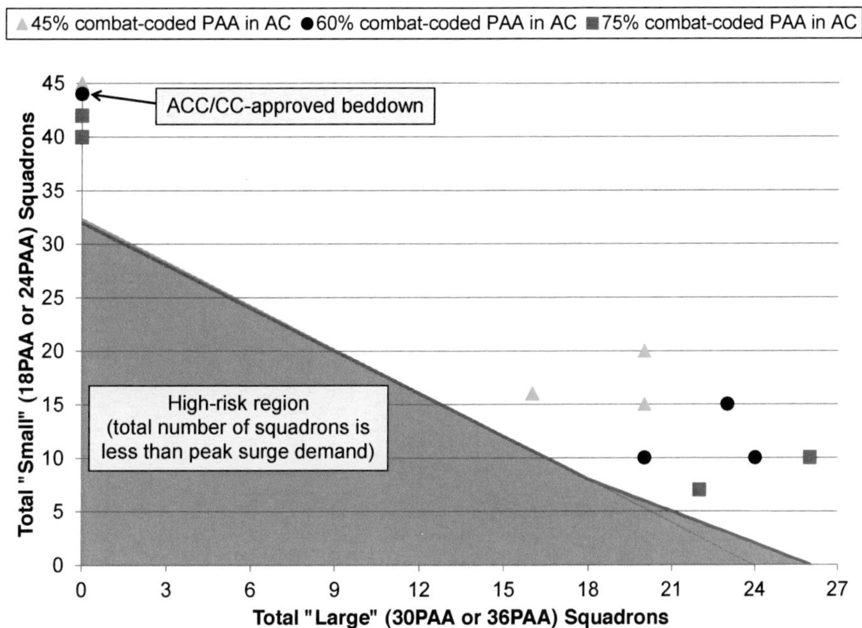

Each marker on the figure corresponds to one paired set of beddown alternatives.[7] The two members of each paired set differ only by the percentage of total AC PAA in CONUS—each member has an equal number of large and small squadrons. For example, beddowns 1G and 1H each have 20 "large" squadrons (in this case, 30 PAA) and 15 "small" ones(in this case, 24 PAA). The red region on this figure corresponds to the range over which the number of squadrons is insufficient to satisfy the peak surge demand. Observe that all beddown alternatives lie outside the red region, meaning all 28 alternatives have sufficient squadrons to satisfy surge squadron requirements.

Most Alternatives Satisfy Rotational Deployment Requirements Within Specified Deploy-to-Dwell Ratios

The primary distinction between surge and steady-state rotational requirements is that deploy-to-dwell considerations limit the number of combat-coded squadrons that are available for rotational deployments at any point in time. This analysis assumes that rotational requirements must be satisfied without exceeding the maximum deploy-to-dwell ratios presented in Table 3. We assume that deployment of a lead UTC counts as a

[7] The marker at 40 "small" squadrons and zero "large" squadrons actually corresponds to six beddowns: 1C, 1D, 2C, 2D, 3C, and 3D.

deployment period for the entire squadron; this is a conservative estimate because not all squadron members would necessarily be deployed—should, for example, a 36 PAA squadron support a deployment of 12 PAA, the average deploy-to-dwell ratio for the individuals in such a squadron would be less than what we present here. Note that these deploy-to-dwell ratios do not imply any specific deployment duration for any unit; rather, they identify the maximum percentage of time that a unit could be deployed, over an indefinite horizon.

Table 3. Maximum Allowable Deploy-to-Dwell Ratio for Rotational Requirements[8]

	NonNNorgeN	NongNNorgeN
C squadrons	C3	C3
C squadrons	C	C3

This analysis assumes that all RC units and all AC units in CONUS are organized as associate units. It is unclear how this organization into associate units would affect the F-35 force presentation model, and thus the maximum allowable deploy-to-dwell ratio in an RC or AC unit. This analysis assumed that the entire unit is available at the host unit's deploy-to-dwell rate. Thus, we assumed that the AC portion of an Active Associate unit was available for rotational deployment at the RC rate. We assumed that the RC portion of a Classic Associate unit was available at the AC rate. This is a conservative assumption for Active Associate units, since it restricts access to the Active Associate AC forces to the lesser availability of RC units. This assumption could be problematic for Classic Associate units, but in the specific sets of deployment requirements that were examined in this analysis, the non-surge and post-surge deployments typically required less than a full squadron's worth of aircraft, and could thus be supported with the AC portion of a Classic Associate unit, provided that the RC portion of such units was not very large.

Alternatively, one could assume that the AC portion of an Active Associate unit was available for deployment at the AC rate. However, this poses difficulties from a force presentation concept. If the AC portion is deployed with the rest of its Active Associate

---

[8] Note that the deploy-to-dwell ratios presented in the post-surge column are consistent with current USAF guidance for periods other than surge. This level of deployment is viewed as the maximum supportable level; however, there are concerns that such a high level of deployment poses challenges to the longer-term sustainability of the force. Thus, based upon consultations with ACC, we modified the deploy-to-dwell ratio in non-surge to allow for less deployment stress on the force during non-surge periods. Note that this increases the requirement for the number of squadrons needed during non-surge periods.

unit, force presentation is maintained as an integral squadron. If the AC portion is available at a different rate than the RC portion, then the AC pilots and maintainers would likely need to be sized to support an entire UTC package(s), with separate RC UTCs providing the remainder of a squadron's designed operational capability statement. In this case, the specific UTCs to be supported by the AC portion would need to be identified. Would the AC support an independent ("lead") or dependent ("follow-on") UTC? Would the force presentation of such AC units assume that AC UTCs deploy with other AC units, leaving the RC remainder to conduct its home-station mission? Or would AC UTCs deploy with *rainbowed* RC units?[9] These force presentation issues were beyond the scope of this analysis, but will require additional study should the USAF decide to utilize associate unit arrangements in the majority of its F-35 squadrons.

Based upon the number of deployed aircraft and number of deployed locations in the ISC non-surge and post-surge scenarios, we identified the minimum number of squadrons necessary to support rotational requirements.[10] These rotational requirements vary slightly based on the squadron size, but the effect is much less than that observed for surge requirements. This is because most non-surge and post-surge operating locations require a relatively small number of aircraft, typically less than the squadron sizes under consideration.

Figure 2 demonstrates how the number of squadrons available in the set of 28 alternative F-35 beddowns compares to these rotational requirements. Again, each marker on the figure corresponds to one paired set of beddown alternatives that differ only with respect to the percent of AC PAA in CONUS. For example, beddowns 1I and 1J are represented by a single point on the figure, since each has 12 AC squadrons and 20 RC squadrons. The solid green region on this figure corresponds to the range over which all non-surge and post-surge demands can be satisfied within the deploy-to-dwell ratios identified in Table 3. The light green region on this figure corresponds to the range over which all rotational requirements can be satisfied within the post-surge deploy-to-dwell ratios.

Observe that most beddown alternatives lie within the solid-green region on this figure. We found that 18 out of 28 total beddown alternatives have sufficient squadrons to satisfy rotational requirements within the less-stressing deploy-to-dwell ratios

---

[9] *Rainbowing* is a deployment strategy used by the RC in which a single deployment requirement is maintained over some duration through the rotation of personnel from multiple RC units.

[10] We note that the maximum total number of squadrons that must be deployed at any point in time during non-surge and post-surge periods is less than the total number of squadrons that must be deployed during surge periods, thus all alternatives have a sufficient total number of squadrons to satisfy all non-surge and post-surge demands.

presented in Table 3, and two additional beddowns could satisfy these requirements if the post-surge deploy-to-dwell ratios were applied during non-surge periods. For beddowns with 75 percent or 60 percent of the combat-coded PAA in the AC, all alternatives except for those with 36 PAA in AC and AFRC squadrons were able to satisfy rotational requirements within the specified deploy-to-dwell ratios. For beddowns with 45 percent of the combat-coded PAA in the AC, only alternatives with 24 PAA in AC and AFRC squadrons were able to satisfy rotational requirements within the deploy-to-dwell ratios; alternatives with 30 PAA in AC and AFRC squadrons and 18 PAA in ANG squadrons were able to satisfy these requirements if the post-surge deploy-to-dwell ratios were applied during non-surge periods. Both increasing the squadron size (i.e., moving down and to the left within the set of triangles, circles, or squares in the figure) and decreasing the fraction of combat-coded PAA in the AC increase the risk that a beddown alternative will not be able to satisfy rotational deployment requirements within the specified deploy-to-dwell ratios.

Figure 2. Ability of Alternative F-35 Beddowns to Satisfy Rotational Deployment Requirements

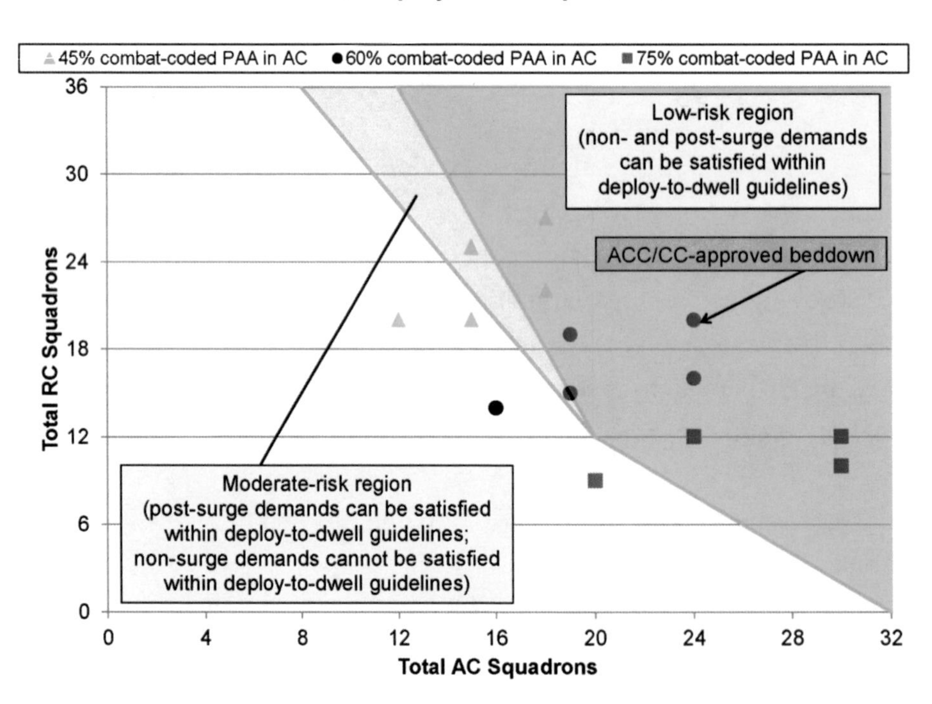

It is important to recognize that, under a different employment construct than is currently envisioned in the ISCs (in which the F-35 is deployed in a manner similar to the F-16), the deployment requirements and associated logistics resource requirements might differ significantly from those presented here. Because the employment of the F-35 is still to be determined by the USAF, potential new concepts such as "many locations with very few F-35s at each location" could significantly change these requirements, and thus the supportability of an F-35 beddown that utilizes large squadron sizes.

Pilot Absorption Requirements Can Be Satisfied Through the Use of Active Associate Units, But This Places Significant Burden on RC Units

Our analyses of pilot absorption capacities were based on a steady-state absorption model that determined whether a beddown alternative satisfied "feasible" absorption conditions.[11] Feasible absorption conditions create enough experienced pilots to generate adequate pilot inventories, using achievable aircraft utilization (UTE) rates, and maintaining acceptable unit experience levels, while enabling pilots to meet specified minimum Ready Aircrew Program (RAP) training requirements across all units in all components.

The historical norm for fighter pilot absorption has been to fill all AC inventory needs, plus all prior-service ANG and AFRC inventory needs, using pilots absorbed primarily in AC units. This has not been possible since the late 1990s, however, because the post–Cold War drawdown took AC force structure below the required levels. Thus, for each of the 28 alternative beddowns, we were provided with a set of three absorption excursions using Active Associations at all RC units, in which differing numbers of AC pilots operate ANG and AFRC airframes in ANG- and AFRC-assigned units (with embedded pilots included in the normal crew ratio authorizations and added pilots assigned over and above normal crew ratios):

Excursion 1:

- **AFRC squadrons:** Two experienced and seven inexperienced AC Combat Mission Ready (CMR) pilots, all embedded.
- **ANG squadrons:** One experienced and three inexperienced AC CMR pilots embedded; five inexperienced AC CMR pilots added.

[11] Absorption capacities measure the number of new pilots that operational units can absorb per year. Operational units absorb new fighter pilots by providing the training, experience, and supervision needed to develop them into combat pilots, instructors, and leaders. Important factors include unit manning and experience levels as well as facilities (e.g., simulators, ranges, and airspace) and aircraft utilization rates.

Excursion 2:

- **AFRC squadrons:** Two experienced and seven inexperienced AC CMR pilots, all embedded.
- **ANG squadrons:** One experienced and three inexperienced AC CMR pilots embedded (no added pilots).

Excursion 3:

- **AFRC squadrons**: Two experienced and seven inexperienced AC CMR pilots embedded per 24 PAA squadron, with embedded pilots increasing proportionally to PAA per squadron for options with PAA > 24.
- **ANG squadrons:** One experienced and three inexperienced AC CMR pilots embedded per 18 PAA squadrons, with embedded pilots increasing proportionally to PAA per squadron for options with PAA > 18.

Similarly, the excursions examined alternatives for Classic Associations (in which ANG and AFRC pilots fly with AC units), which were assumed to exist for every CONUS-based AC unit.

The analysis found that achieving feasible absorption conditions will require both a change in the burden historically borne by RC units and additional resources allowing AC units to overfly RAP minimums. Only one of the analyzed excursions produced pilot inventories that approached the required levels, and all excursions tended to impose a disproportionate share of the absorption burden on the ANG and AFRC units. The first absorption excursion required ANG unit UTE rates that are two to three sorties per airframe per month (15 to 23 percent) greater than the AC UTE for many beddown alternatives, and forced ANG unit experience levels to drop below 60 percent for several beddown alternatives. Within the first absorption excursion, six of the 28 beddown alternatives, each of which assumed 45 percent of combat-coded PAA in the AC and 24 PAA per ANG squadron, did not generate the required pilot inventories, with shortfalls ranging from 31 to 98 pilots. These pilot shortfalls can be eliminated, if AC units are allowed to overfly the RAP minimums (the corresponding AC overfly for these beddown alternatives ranged between 2.4 and 7.5 percent above the RAP minimums).

We found that squadron size and AC/RC mix affected experience levels in RC units; i.e., RC experience level increases with squadron size and with the percentage of aircraft in the AC. The RC UTE requirement to meet pilot absorption decreases as squadron size increases; this requirement was not significantly affected by the AC/RC mix. Because the number of associated AC pilots per unit does not vary with RC squadron size in the first excursion, the inexperienced AC pilots have a lesser effect on the overall experience level for a larger RC squadron, and the increased flying needed to support the AC pilots is distributed over a larger number of aircraft in larger RC squadrons. As the percent of aircraft in the AC increases, more new AC pilots are absorbed each year, which in turn generates a larger pool of AC pilots who eventually depart the AC as experienced pilots and affiliate with RC units, decreasing the RC units' requirement to train their own inexperienced non-prior-service pilots.

The AC UTE requirement decreases as the percentage of total aircraft in AC increases; this requirement was not significantly affected by squadron size. This is because the total AC pilot inventory requirement includes a large number of pilots who are not in F-35 operational units, but who are needed for other missions, such as test and training squadrons, or staff positions. This requirement for AC pilots outside of the F-35 operational units was assumed to be constant across all beddown alternatives; thus, alternatives with less aircraft in the AC have fewer AC units through which to absorb the total pilot requirement, whereas alternatives with more aircraft in the AC have a broader base of AC units through which the non-operational units' fighter pilot requirements can be absorbed.

Increasing Squadron Sizes Reduces Pilot Absorption Flying Costs

Under the first absorption excursion (given the set of UTE requirements identified for each of the 28 alternative beddowns), we identified the annual cost associated with generating the required number of sorties (assuming an average sortie duration of 1.4 flying hours per sortie, and a cost of $18,025 per flying hour).[12] Figure 3 presents the annual pilot absorption costs associated with each of the 28 beddown alternatives, with the value for each alternative presented as the percentage difference between its cost and the cost of the baseline ACC/CC-approved beddown. Each marker on the figure again corresponds to one paired set of beddown alternatives—each member of the set has an equal number of RC and AC squadrons, they differ only in the percent of total AC PAA in CONUS (which did not have a significant impact on the costs presented here).

---

[12] The Air Force Cost Analysis Agency (AFCAA) provided us with an F-35A steady state cost per flying hour (CPFH) in base year 2012 dollars. "Steady state" is defined here as the average cost during the period with the maximum number of PAA, which for the F-35A is fiscal years (FYs) 2036–2040. This factor includes cost growth above inflation, and comprises costs for fuel ($6,604), consumables ($1,793) and depot-level repairables ($9,628).

Figure 3. Annual Pilot Absorption Flying Costs, by Beddown Alternative

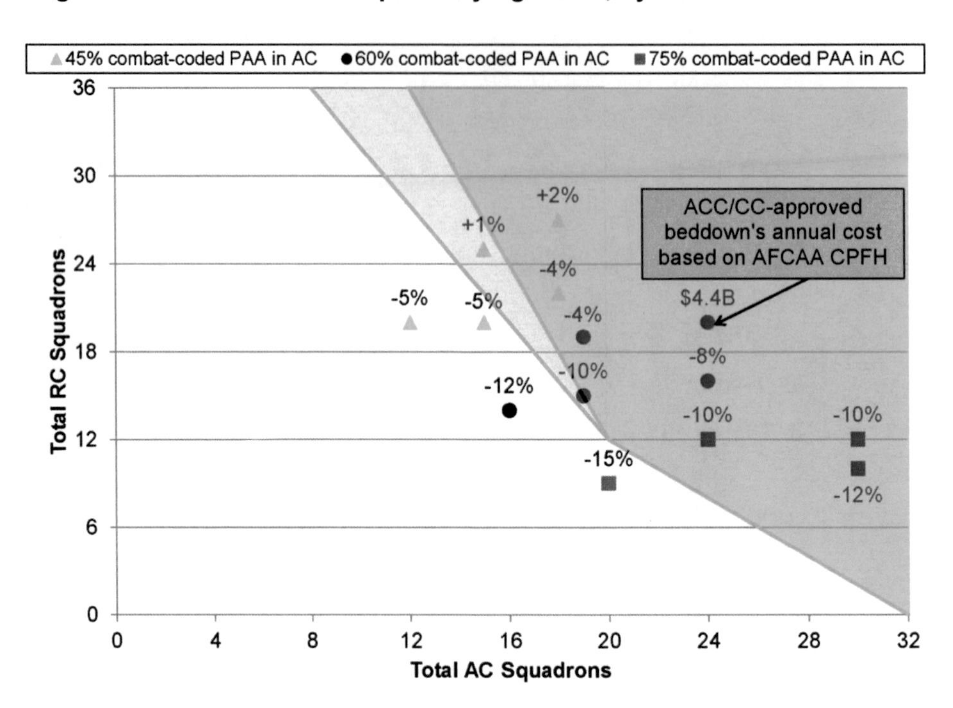

As the fraction of combat-coded PAA in the AC is held constant (i.e., within the set of circles, squares, or triangles in the figure), increasing the squadron size (i.e., moving down and to the left on the figure, with fewer squadrons) can significantly reduce the annual pilot absorption flying cost. Observe that the ACC/CC-approved beddown has an annual pilot absorption flying cost of $4.4 billion. Within the alternative that maintains 60 percent of the combat-coded PAA in the AC, increasing AC and AFRC squadron size to 30 PAA while maintaining 18 PAA per ANG squadron can reduce these costs four percent relative to the ACC/CC-approved beddown, while increasing ANG squadron size to 24 PAA and maintaining 24 PAA per AC and AFRC squadron could reduce these costs by eight percent. Increasing both AC and AFRC squadrons to 30 PAA and ANG squadrons to 24 PAA would reduce these costs by 10 percent, while a further increase to 36 PAA in the AC and AFRC could reduce these costs by 12 percent.

As the squadron size is held constant, increasing the fraction of combat-coded PAA in the AC (e.g., comparing the marker farthest to the left for each colored set of markers) also generates cost reductions. When compared to the ACC/CC-approved beddown's $4.4 billion in annual pilot absorption flying costs, there are many alternative beddowns that satisfy all deployment requirements and reduce this cost by 10 percent or more.

Finding Feasible Pilot Absorption Results

Because none of the absorption excursions that we were asked to analyze generated a feasible absorption condition that also maintained the RC's higher experience levels and lower UTE rates, relative to the AC, we performed additional analysis to search for a set of assumptions that would satisfy all of these objectives. These additional provisions lead to a relatively simple feasible absorption condition:

- **AFRC squadrons**: Two experienced and seven inexperienced AC CMR pilots embedded per 24 PAA squadron, with embedded pilots increasing proportionally to PAA per squadron for options with PAA > 24 (identical to excursion 3 above).
- **ANG squadrons**: Two experienced and seven inexperienced AC CMR pilots embedded per 24 PAA squadron, with embedded pilots decreasing proportionally to PAA per squadron for options with PAA = 18 (similar to AFRC).

If a 10 percent overfly of RAP minimums is enforced at all AC units, all beddown alternatives that have at least 60 percent of the F-35 combat-coded PAA in the AC remain feasible, regardless of whether ANG units contain one or two experienced AC supervisors.[13] AC UTE rates compare favorably with the UTE rates required in ANG and AFRC units in the sense that in all but one case they are at least as large or larger. The required UTE rates range from 13.4 to 14.8 for AC units and from 12.2 to 14.0 in RC units.[14] Moreover, in all beddowns with 45 or 60 percent of the combat-coded PAA in the AC, the total annual flying-hour requirement is less for this feasible excursion than for the first absorption excursion.[15] If decisionmakers desire to maintain a more traditional UTE relationship among the components' units, they will need to accept a higher allowable overfly of RAP minimums at AC units.

This is a single feasible result that was found once the problem's bounds were adequately relaxed. Additional analysis could better characterize the set of feasible solutions and identify alternatives that are optimal in meaningful contexts for Air Force leaders in all components.

---

[13] Beddown alternatives with 45 percent of the combat-coded PAA in the AC require an overfly of RAP minimums by 10 to 16 percent at AC units to generate sufficient pilot inventories.

[14] Again, UTE rates are measured as sorties per PAA per month.

[15] The feasible excursion requires between 0.4 and 8.3 percent fewer FH than the first excursion for all beddowns with either 45 or 60 percent of combat-coded PAA in the AC. For corresponding beddowns with 75 percent of combat-coded PAA in the AC, the feasible excursion's pilot absorption flying cost is between 0.6 and 2.8 percent greater than the first excursion.

Increasing Squadron Sizes Reduces Maintenance Manpower Requirements

We found that, for combat-coded aircraft, the required maintenance manpower per PAA decreases as the number of PAA per squadron increases. We estimated that a squadron of 36 PAA could be supported by 26 percent fewer maintenance positions per PAA than could a single squadron of 18 PAA. Furthermore, assigning multiple squadrons to a single wing can generate additional savings beyond those generated by the squadron size effect. Our analysis suggests that a wing of three 36 PAA squadrons requires 6 percent fewer maintenance positions per PAA than a single squadron of 36 PAA.

Figure 4 presents the total annual manpower costs associated with each of the 28 beddown alternatives, with the value for each alternative presented as the percentage difference between its cost and the cost of the baseline ACC/CC-approved beddown. Each marker on the figure again corresponds to one paired set of beddown alternatives—each member of the set has an equal number of RC and AC squadrons, and differ only in the percentage of total AC PAA in CONUS.

Figure 4. Total Annual Maintenance Manpower Costs, by Beddown Alternative

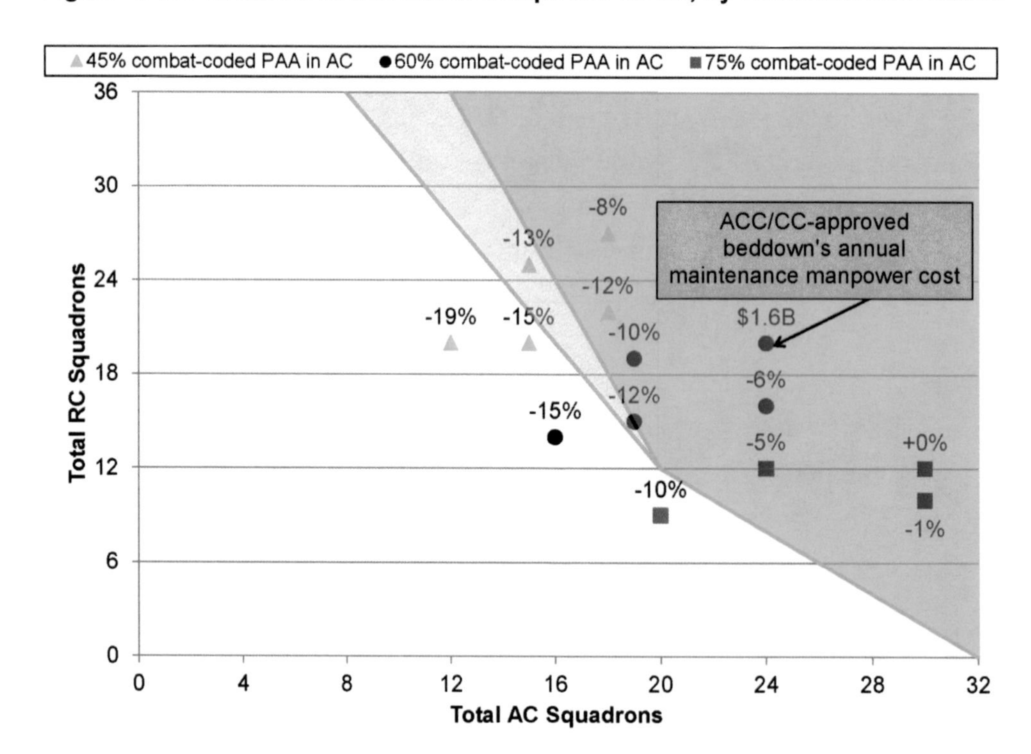

As the fraction of combat-coded PAA in the AC is held constant (i.e., within the set of circles, squares, or triangles in the figure), increasing squadron size (i.e., moving down and to the left on the figure) can significantly reduce the overall maintenance manpower cost. This is consistent with the manpower economies of scale discussed above. However, as the squadron size is held constant, increasing the fraction of combat-coded PAA in the AC (e.g., comparing the marker farthest to the left for each colored set of markers) increases the overall cost. This occurs because the RC is able to make use of part-time maintainers, who are much less expensive in a nondeployed steady-state role than AC maintainers.

The analysis presented thus far (which we will term the first alternative) has assumed that Active Associate units utilize only RC maintenance manpower, with the AC providing no maintenance manpower to the associate units. We also considered two other strategies that place AC maintenance manpower in the Active Associate unit. Under the second alternative, the AC would provide the maintenance manpower necessary to support the increased home-station flying caused by AC pilots.[16] Under the third alternative, in addition to increasing pilot absorption capabilities, the AC manpower is used to generate an increased deployment capability: We identify the AC manpower necessary to support an entire set of UTCs, position these UTCs within Active Associate units, and make them available at the deploy-to-dwell ratios assumed for the AC. Under both the second and third alternatives, RC full-time manpower is reduced as AC maintenance manpower is added to the unit. However, because the typical AC maintainer would be expected to be less-experienced and less-productive than the typical RC maintainer, these positions were not traded on a one-for-one basis: an equivalency factor of approximately 1.44 AC maintainers per full-time RC maintainer was assumed. Across the three alternatives considered, the total annual cost differs by no more than 1.3 percent. Said differently, the third alternative provides more deployment capability at essentially the same total cost. However, the number of AC maintainers required at Active Associate units varies significantly. For those beddowns that maintain 60 percent of combat-coded PAA in the AC (as in the ACC/CC-approved beddown 2A), the first and second alternatives can satisfy the AC pilot absorption requirements with between zero and 168 total AC maintenance positions at Active Associate units, while the third alternative also provides an increased steady-state deployment capability through the use of between 980 and 1,400 total AC maintenance positions at Active Associate units.

[16] Note that this increased home-station flying was incorporated into the requirements for the first alternative, but it was not necessary to separate the home-station flying into different segments because RC manpower were performing all maintenance.

Because we found little difference between the manpower composition alternatives with respect to total annual maintenance manpower costs, the key tradeoff to be considered when evaluating these alternatives is the increase in deployment capability that can be achieved under the third alternative versus the increased AC maintenance manpower requirements at Active Associate units.

Increasing Squadron Size Reduces Support Equipment Procurement Costs

Given the SE procurement costs for each squadron and wing under consideration, we can similarly identify the total SE procurement cost across each of the 28 alternative F-35 beddowns. Figure 5 presents the total SE procurement costs associated with each of the 28 beddown alternatives, with the value for each alternative presented as the percentage difference between its cost and the cost of the baseline ACC/CC-approved beddown. As with maintenance manpower, each marker on the figure corresponds to one paired set of beddown alternatives, because we observed that the percentage of AC PAA assigned to CONUS locations had little impact on these costs.

Figure 5. Total SE Procurement Costs, by Beddown Alternative

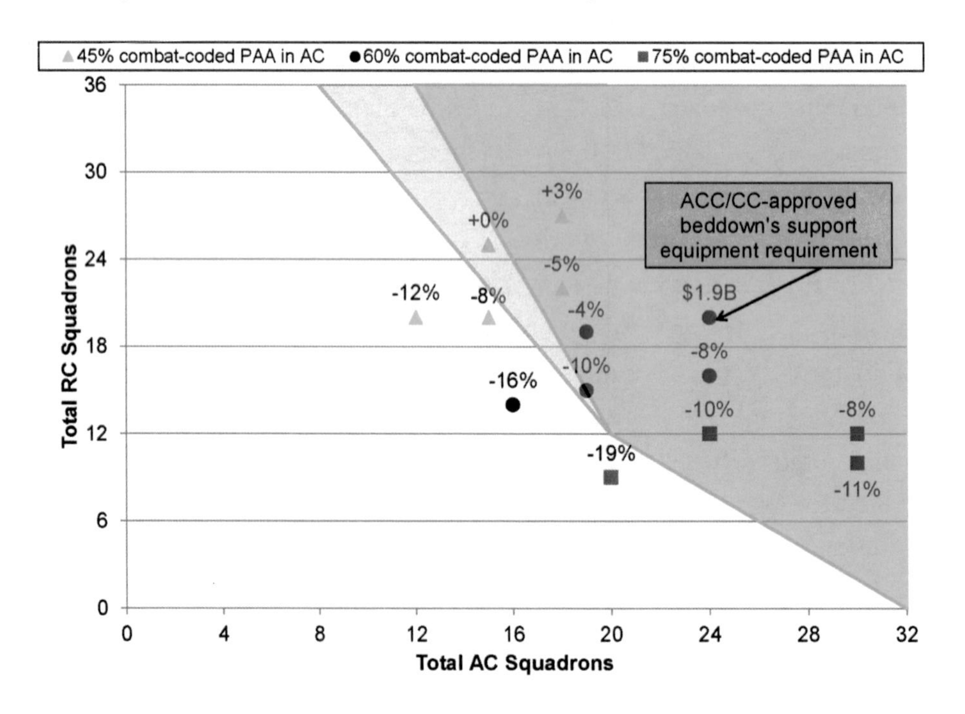

As the fraction of combat-coded PAA in the AC is held constant, increasing squadron size can significantly reduce overall SE procurement cost because economies of scale also exist for SE requirements. Furthermore, as squadron size is held constant, increasing the fraction of combat-coded PAA in the AC also decreases the overall cost. This occurs because the ANG is limited to smaller squadron sizes, and when the fraction of total PAA in the AC is increased, fewer PAA are assigned to the smaller PAA ANG squadrons.

Little Additional Infrastructure Capacity at Existing F-16 and A-10 Bases Would Be Needed to Support the F-35

Our analysis considered infrastructure capacity at the set of existing F-16 and A-10 bases, across six resource categories for the F-35.[17] As noted in Figure 6, for runway and ramp, no new capacity is needed under all beddown alternatives—all F-35 requirements can be satisfied with existing infrastructure. The other resource categories (squadron operations/Aircraft Maintenance Unit, ammunition storage, corrosion control and maintenance) did require some additional capacity (denoted by the cross-hatched areas in the figure); however, in most cases, these requirements are relatively small.[18]

The beddown alternatives also exhibit some cost reductions associated with consolidation to fewer bases. Larger squadron sizes reduce annualized facilities costs, while increasing the percentage of aircraft in the AC reduces facility costs, for this set of infrastructure resources.

---

[17] This is not an exhaustive list of additional infrastructure required at a current F-16 or A-10 base in order for the base to support F-35 operations. As an example, based on the increased security classification requirements for fifth-generation fighter aircraft, increased costs would be necessary to support a higher level of classification for communications lines, sensitive compartmented information facilities, etc.

[18] Note that this is based on analysis of raw square footage data from an Office of the Secretary of Defense database (the Facilities Program Requirements Suite) and does not address the condition or adequacy of current facilities and infrastructure.

Figure 6. F-35 Infrastructure Requirements, by Beddown Alternative

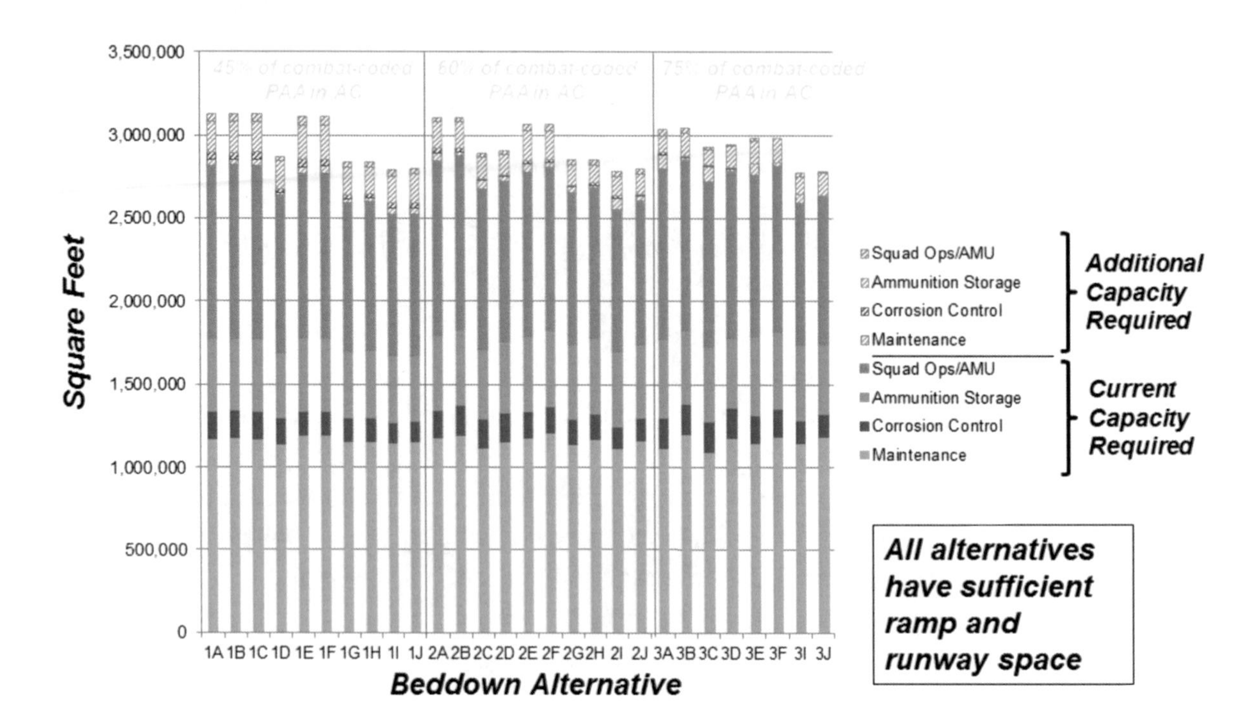

Leader Development Is More Affected by the Assignment Policy Used Than Squadron Size or AC/RC Mix

The F-35 beddown alternatives would substantially alter the numbers of PAA and units in the AC, ANG, and AFRC—and, consequently, the numbers of jobs such as squadron commander, group commander, and wing commander that are regarded as key developmental experiences. Hence, Air Force decisionmakers asked whether some beddown alternatives would endanger the development of future senior leaders. We assessed the Air Force's capacity for developing fighter pilots in the AC under the beddown alternatives. Different developmental goals call for somewhat different career paths. For this analysis we focused on well-prepared candidates for selection as general officers along one of two broad developmental paths:

- Joint warfare roles as, e.g., combatant command commanders, vice commanders, operators (J-3s), and planners (J-5s)

- Organize/train/equip roles as, e.g., chief, vice chief, or assistant chiefs of staff at Air Force headquarters or headquarters in the major commands.[19]

We postulated two other career paths for fighter pilots who could rise to colonel, one oriented toward organize/train/equip roles and one for general purposes, essentially filling in O-6 jobs as necessary. We also considered one path to represent officers who advance into the field grades but not to colonel, and one for officers who leave before promotion to major, both paths filling jobs in their grades that the other four paths do not fill completely.

To help guide and control the flows of fighter pilots through the jobs and career paths identified, we used historical data to derive the steady-state experience and grade profile for fighter pilots. We used optimization and simulation models to estimate the AC's capacity to develop fighter pilots. The optimization model maximizes the flows through the two general officer paths, its results represent upper limits on the numbers of officers who could be channeled through the various career paths. The simulation model estimates what would happen in a less carefully managed environment, which represents the progression of simulated individuals through jobs and grades in a manner that is more similar to how the Air Force manages actual assignment and promotion processes.

Figure 7 displays results from the optimization and simulation models, reflecting the total numbers of fighter-pilot colonels graduating with the combinations of experience deemed mandatory for the general officer paths, across all 28 beddown alternatives. The tops of the bars reflect the optimization's results and the bottoms the simulation's results. The simulation's annual production of fighter pilots with the experiences marked as mandatory on the two paths varies from 6.2 to 7.7 per year, a difference of 1.5 or about 24 percent. The flow optimization produces from 10.6 to 14.2 pilots per year, a difference of 3.6 or about 34 percent across these beddown alternatives. Given that five to seven fighter pilots per year have been promoted to general officer during recent years, these results suggest that the USAF will be somewhat constrained with respect to fighter pilot leadership development. To allow for a larger pool of candidates with the preferred characteristics, the USAF needs to be deliberate with its leadership development during the change from legacy fighter/attack aircraft to the F-35.

---

[19] These developmental approaches had been employed in earlier, unpublished analysis for the Air Force Chief of Staff and the Secretary of the Air Force. If desired, additional research could allow distinct paths to be tailored to grow fighter pilots with more specific areas of expertise—e.g., in acquisition, logistics, human resources, or political-military affairs.

Figure 7. F-35 Beddown Alternatives Affect Leader Development by Less Than Assignment Policy

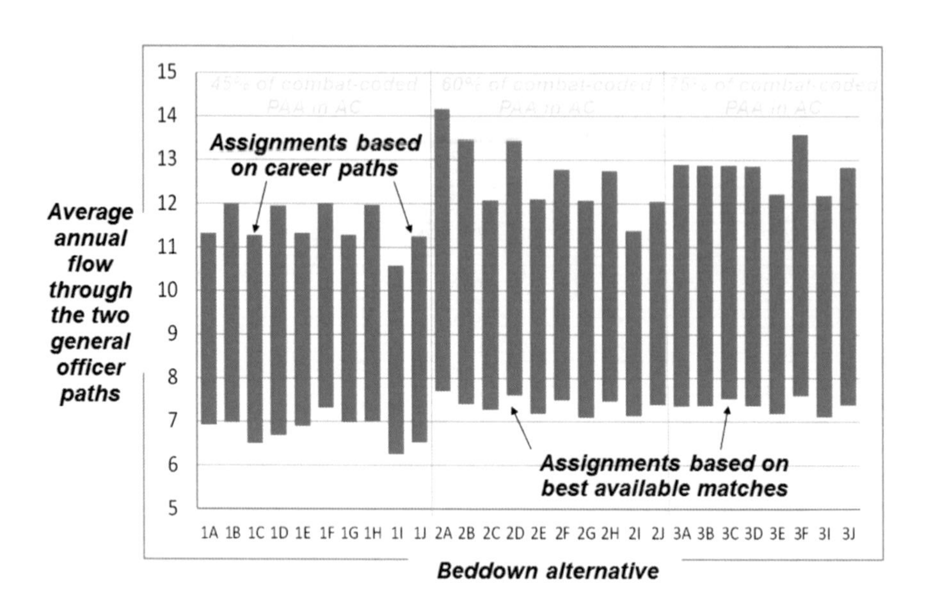

Note that the spans between the bottoms and tops of these bars range between 60 percent and 84 percent. That is, using planned career paths to guide assignments (i.e., building officers' combinations of experiences consistent with their perceived potential for future leadership roles) would have more effect than the F-35 beddown alternatives on the AC's production of candidates for leadership for these roles at the general officer level. In other words, the analysis found that leader development was more affected by the assignment policy used than by squadron size or the percentage of combat-coded PAA in the AC.

In total, we concluded that the F-35 beddown alternatives would have some effect on the AC's capacity for producing future senior leaders with targeted combinations of experience, but none of the alternatives with at least 60 percent of the combat-coded PAA in the AC would jeopardize its ability to produce at least as many well-qualified candidates as have actually been promoted to general officer during recent years.

The Way Forward

The findings from this analysis can be used to inform many issues that are within the purview of other USAF analyses and decision processes, including the Total Force Integration Roundtable's discussion of Associate Unit Force Presentation, the Directorate of Strategic Planning, Office of the Deputy Chief of Staff for Strategic Plans and Programs, Headquarters USAF (AF/A8X)'s Multi-Role Fighter Phase II Force Composition Analysis, and the Strategic Basing Process performed by the Office of the

Assistant Secretary of the Air Force for Installations, Environment, and Logistics (SAF/IE) and the Office of the Deputy Chief of Staff for Strategic Plans and Programs, Headquarters U.S. Air Force (USAF/A8). In particular, these findings can help determine how F-35 associate units should be composed and resourced in order to meet the requirements of increased pilot absorption and (potentially) increased deployment capability.